EXTREME WEATHER

HAIL

by Liza N. Burby

The Rosen Publishing Group's
PowerKids Press™
New York

Published in 1999 by The Rosen Publishing Group, Inc.
29 East 21st Street, New York, NY 10010

First Edition

Book Design: Resa Listort

Photo Credits: p. 4 © Kuni/AP Photo; p. 7 © Richard Hamilton Smith/Corbis; pp. 8, 16 © Ron Heflin/AP Photo; p. 11 © Eric Gilbert, Papilio/Corbis; pp.12–13 © Daphne Kinzler, Frank Lane Picture Agency/Corbis; p. 15 © Annie Griffiths/Corbis; p. 19 © Nick Ut/AP Photo; pp. 20–21 © Annie Griffiths Belt/Corbis.

Burby, Liza N.
    Hail / by Liza N. Burby.
        p.   cm. — (Extreme weather)
    Includes index.
    Summary: Describes hailstorms, how they begin, how hailstorms are formed, the damage they can cause,
and safety precautions to observe during such storms.
    ISBN 0-8239-5293-2
    1. Hail—Juvenile literature. [1. Hail.]
    I. Title. II. Series: Burby, Liza N. Extreme weather.
QC929.H15B87 1998
551.57'87—dc21   98-7566
                                                        CIP
                                                         AC

Manufactured in the United States of America

Contents

1   What Is Hail?   5

2   When and Where Do Hailstorms Happen?   6

3   What Does Hail Look Like?   9

4   How Does It Start?   10

5   How Does Hail Behave?   12

6   Why Is Hail Dangerous?   14

7   Strange Things Hail Does   17

8   Hail in History   18

9   Safety During a Hailstorm   21

10   Different Kinds of Frozen Precipitation   22

   Glossary   23

   Index   24

What Is Hail?

Thunder booms and **lightning** (LYT-ning) flashes. Rain pours from the sky. Suddenly, the rain feels lumpy. It looks as if little rocks are falling from the sky! But these aren't rocks—they're actually pieces of ice called **hail** (HAYL). Each piece is called a hailstone. Hailstones bounce off cars and sidewalks. They make a clicking sound as they fall, like thousands of marbles bumping into one another. Hail looks funny, but actually it can be dangerous. Farmers in the Midwest call hail the "white **plague** (PLAYG)."

It can really hurt when you get hit by hail!

When and Where Do Hailstorms Happen?

Hailstorms happen during thunderstorms, which usually occur during the summer months. They usually start in the middle or late afternoon and last about fifteen minutes or longer. In the United States, the area that has about eight to ten hailstorms a year is called Hail Alley. This area is made up of a line of states stretching from southeastern Wyoming into northern Colorado and into western Nebraska. Hail Alley is second in the world for its number of hailstorms. The first is the country of Kenya in Africa. There, hailstorms occur about 132 days a year.

Hail Alley spreads through the Southwest and Midwest of the United States. ▶

What Does Hail Look Like?

People often think that hailstones are round. But hailstones actually come in different shapes. Some are shaped like eggs and some have uneven edges. The uneven hailstones are **clusters** (KLUS-terz) of smaller stones that have frozen together. Hailstones also come in different sizes, and people often measure them to see how big they are. Some are as small as a pea, others are the size of a walnut, and still others can be as large as a softball. If you cut a large hailstone in half, you will see it has many layers. Hailstones can have as many as 25 layers.

This woman from Texas shows three large hailstones that she found in her backyard.

How Does It Start?

Hail will usually happen during a spring or summer thunderstorm. It comes from large, dark clouds called thunderheads or **cumulonimbus** (kyoo-myoo-loh-NIM-bus) clouds. Hail starts out as tiny water drops in a cloud. Inside the cloud, air moves around and around. Water droplets move around with the air. As the droplets move to the top of the cloud where it is very cold, they freeze. And when they move toward the bottom of the cloud, they start to melt. Traveling up again, the droplets freeze. The droplets gather more layers as they continue this **cycle** (SY-kul). Soon they become too heavy to do anything but fall to the ground as hailstones.

When a hailstrom occurs, the sky can turn black as if it were nighttime. ▶

How Does Hail Behave?

Hailstones can fall as fast as 140 miles per hour. Big hailstones of about two to three inches in **diameter** (dy-AM-eh-ter) may fall up to 100 miles per hour. When hailstones are large, you can hear a rumble when they fall. Some people believe they can hear the

hailstones click together in the sky before they hit the ground. When you hear a clatter outside, you can always tell it is hailing.

Hailstones can pile up during a hailstorm and form a slippery, frozen mess.

Why Is Hail Dangerous?

Hailstones are hard. When they are large, they are also dangerous. People and animals can be hurt or even killed by falling hailstones. On April 30, 1888, baseball-sized hailstones killed 246 people in India. On June 19, 1932, 200 people died in a hailstorm that lasted two hours.

Each year, about half a billion dollars in crops, farm animals, and property in the United States are destroyed by hail. Farmers can lose all of their crops to hail. It flattens wheat, ruins corn, and cuts melons, tomatoes, and flowers. It can also dent cars and planes and break windows.

Some hailstorms are so strong they can kill even large plants such as these sunflowers. ▶

Strange Things Hail Does

Sometimes when little hailstones get caught in their cycle in the clouds, very strong air **currents** (KUR-entz) pull other things along into the cloud for the ride. Sometimes these things are small animals like frogs and turtles. These animals can get stuck inside a hailstone! People have seen hailstones with frogs and turtles inside them. A hailstone about the size of an egg once fell on a farm in Essen, Germany. Inside it was a small fish!

The bigger the hailstones, the more damage they can cause. These cars were dented by hailstones that were the size of baseballs.

Hail In History

Cheyenne, Wyoming, is known as the hailstorm capital of the United States because it has so many hailstorms. On August 3, 1985, the people of Cheyenne were surprised when rain turned to hail. When the storm was over, twelve people had been killed and 70 more were hurt. Even though it was summertime, there were five-foot **drifts** (DRIFTS) of hail in the city!

In September 1970, hailstones bigger than grapefruit fell on Coffeyville, Kansas. The largest hailstone ever recorded landed there. It weighed about two pounds and was seventeen inches in diameter! That's bigger than a basketball!

Sometimes the only way to look for people stuck in a storm is by helicopter, like this one in California. ▶

Safety During a Hailstorm

If a light rainstorm changes to heavy rain and then hail, be careful. Even a hailstorm with small-sized hailstones can quickly become dangerous and the stones can grow larger. Small hailstones can sting, and baseball-sized chunks can definitely hurt you. Larger hailstones often fall close to where a tornado is starting. Stay indoors and listen to weather reports. When a hailstorm is over, you can go outside and break a hailstone in half. Then you will see its layers of ice.

If you see clouds that look like this, you should find shelter quickly.

Different Kinds of Frozen Precipitation

There are a few kinds of frozen **precipitation** (pre-SIP-ih-TAY-shun) besides hail. Soft hail or snow grains are called **graupel** (GRAW-pul). Graupel is frozen cloud droplets. It is soft when it lands on the ground and then it flattens out. Graupel usually happens during bad lightning storms and often hail will form around it.

Sleet is another form of frozen precipitation that happens when rain freezes or when slightly melted snowflakes freeze. When raindrops fall into cold air, sleet will happen. If a cold cloud passes over a mountaintop, cloud droplets freeze on anything they touch. This is called **rime** (RYM). Rime forms on trees and other objects on mountaintops.

Hail is an interesting part of nature and our weather. But check it out from the safety of your home.

Glossary

cluster (KLUS-ter) A group.

cumulonimbus (kyoo-myoo-loh-NIM-bus) Having to do with a thunder cloud.

current (KUR-ent) The flow of water or air in a certain direction.

cycle (SY-kul) The period of time in which an event happens again and again.

diameter (dy-AM-eh-ter) The distance across the center of a circle or sphere.

drift (DRIFT) Mounds of snow or hail that have been piled up by the wind.

graupel (GRAW-pul) Soft hail.

hail (HAYL) Small pieces of ice that sometimes fall during a thunderstorm.

lightning (LYT-ning) A flash of electricity seen as light in the sky that often happens before or during a storm.

plague (PLAYG) Something, like a disease, that spreads quickly.

precipitation (pre-SIP-ih-TAY-shun) Rain, snow, or any moisture that falls from the sky.

rime (RYM) Frost.

Index

C
clusters, 9
cumulonimbus, 10
currents, 17
cycle, 10

D
diameter, 12, 18
drifts, 18

G
graupel, 22

H
Hail Alley, 6
hailstones, 5, 9, 10,
    12–13, 14, 17, 18, 21
hailstorms, 6, 14, 18, 21

L
lightning, 5, 22

P
plague, 5
precipitation, 22

R
rime, 22

S
sleet, 22
summer, 6, 10

T
thunder, 5
thunderstorms, 6, 10
tornadoes, 21